ISBN 978-1-5280-2444-0
PIBN 10899939

1 MONTH OF
FREE
READING

at

www.ForgottenBooks.com

By purchasing this book you are eligible for one month membership to ForgottenBooks.com, giving you unlimited access to our entire collection of over 1,000,000 titles via our web site and mobile apps.

To claim your free month visit:

www.forgottenbooks.com/free899939

English
Français
Deutsche
Italiano
Español
Português

www.forgottenbooks.com

Mythology Photography **Fiction**
Fishing Christianity **Art** Cooking
Essays Buddhism Freemasonry
Medicine **Biology** Music **Ancient
Egypt** Evolution Carpentry Physics
Dance Geology **Mathematics** Fitness
Shakespeare **Folklore** Yoga Marketing
Confidence Immortality Biographies
Poetry **Psychology** Witchcraft
Electronics Chemistry History **Law**
Accounting **Philosophy** Anthropology
Alchemy Drama Quantum Mechanics
Atheism Sexual Health **Ancient History**
Entrepreneurship Languages Sport
Paleontology Needlework Islam
Metaphysics Investment Archaeology
Parenting Statistics Criminology
Motivational

Historic, archived document

Do not assume content reflects current
scientific knowledge, policies, or practices

ERRORS IN THE WEIGHT OF PRINT BUTTER

THEIR CAUSES AND PREVENTION

H. RUNKEL

Assistant Chemist, Bureau of Chemistry

AND

H. M. ROESER

Assistant Physicist, Bureau of Standards
U. S. Department of Commerce

CONTENTS

CIRCULAR 95, OFFICE OF THE SECRETARY
UNITED STATES DEPARTMENT OF AGRICULTURE

Contribution from the Bureau of Chemistry, Carl L. Alsberg, Chief

Washington, D. C.   April, 1918

WASHINGTON : GOVERNMENT PRINTING OFFICE : 1918

ERRORS IN THE WEIGHT OF PRINT BUTTER: THEIR CAUSES AND PREVENTION.

REASON FOR THE INVESTIGATION.

During a recent investigation by the Bureau of Chemistry, in cooperation with the Division of Weights and Measures of the Bureau of Standards,[1] about 400 buttermaking establishments, well distributed over the dairy sections of the United States, were visited, and some 13,000 packages of print butter, before they had left the hands of the manufacturers, were weighed. In many instances, creamery men, actuated by a desire to avoid conflict with officials for violating the Federal net weight regulation, were found to be giving considerable overweight. Inquiry among practical creamery men and a comparison of the different methods and processes of printing convinced the investigators that but a vague conception of the requirements of the Federal net weight law and of the possibilities for the accurate subdividing and printing of butter with the machinery and utensils now obtainable exists. This bulletin has been prepared to show the responsibility of butter manufacturers in complying with the requirements of the Federal net weight law, and to indicate a procedure which the investigators believe will materially reduce the inaccuracy and lack of uniformity due to the present methods of printing butter.

Butter, when shipped in interstate or foreign commerce, or made or sold in the District of Columbia or the Territorial possessions of the United States, is misbranded if in package form the quantity of the contents be not plainly and conspicuously marked on the outside of the package. The net contents should be stated in terms of weight. Packages containing one pound or over should be marked in terms of pounds, and quantities of less than one pound may be stated in simple fractions of a pound or in ounces. The regulation under the

[1] Acknowledgment is made of the helpful comments and suggestions of several members of the Bureau of Standards, of the Department of Commerce; the Bureau of Animal Industry and the Bureau of Chemistry, of the Department of Agriculture; Dr. E. S. Guthrie, of the Department of Dairy Industry, Cornell University; and for the permission of various manufacturers to reproduce their printing devices.

35547°—18-

3

net weight amendment (Regulation 29, amended) is in part as follows:

(*b*) The quantity of the contents so marked shall be the amount of food in the package.

(*h*) The quantity of the contents may be stated in terms of minimum weight, minimum measure, or minimum count, for example, "minimum weight, 10 oz.," "minimum volume, 1 gallon," or "not less than 4 oz.;" but in such case the statement must approximate the actual quantity, and there shall be no tolerance below the stated minimum.

(*i*) The following tolerances and variations from the quantity of the contents marked on the package shall be allowed:

(1) Discrepancies due exclusively to errors in weighing, measuring, or counting which occur in packing conducted in compliance with good commercial practice.

(2)   * * *   (refers to variation of bottles).

(3) Discrepancies in weight or measure, due exclusively to differences in atmospheric conditions in various places, and which unavoidably result from the ordinary and customary exposure of the packages to evaporation or to the absorption of water.

Discrepancies under classes (1) and (2) of this paragraph shall be as often above as below the marked quantity. The reasonableness of discrepancies under class (3) of this paragraph will be determined on the facts in each case.

(*j*) A package containing two avoirdupois ounces of food or less is "small" and shall be exempt from marking in terms of weight.

The work of establishing tolerances for use by those charged with the enforcement of this act is still in progress. In the meantime, certain States, taking up the spirit of the regulation, have established for intrastate commerce allowable variations from the declared weight of butter and other commodities, some examples of which are as follows:

Connecticut allows $\frac{1}{4}$ ounce in deficiency on 16-ounce prints.

New York. "The maximum variation allowed on a pound print is three-eights of an ounce on an individual print, provided that the average error of twelve prints, taken at random, shall not be over one-fourth of an ounce per pound. The maximum variation allowed on a two-pound print is one-half ounce, provided that the shortage on twelve prints, taken at random, shall not be over one-fourth of an ounce per pound. The maximum variation allowed on a two-pound print is one-half ounce, provided that the shortage on twelve prints, taken at random, be not more than three-eighths of an ounce for two pounds."

Pennsylvania allows a variation of $\frac{1}{16}$ ounce on 1-pound prints.

The men collecting data for the investigation in the States where tolerances are enforced found that many buttermakers were unaware of the existence of the tolerances. Of more than 200 buttermen who were asked to state the difference between the heaviest and the lightest print made by their printer in the course of a day's run, all but 42 either had no idea or thought there was no appreciable variation, while the average of the 42 guesses made was 0.33 ounce. Fifty prints selected at random were weighed at each of some 250 plants. Actually the average difference between the heaviest and the lightest of the prints selected in this manner was found to be 0.56 ounce, nearly twice as great as the average estimated amount. The butter-maker evidently believes that but small errors result from casual control, while in fact the clear and simple requirements of the law can be met only by a careful control of the printing process.

TYPES OF ERRORS MADE.

Two types of errors may be made by manufacturers of 16-ounce butter prints, or prints of any size:

First, if each finished print is weighed with sufficient care and accuracy it will be found that, under ordinary conditions, some prints are slightly over, and some are slightly under, the proper weight. *The difference of single prints from 16 ounces*, whether over or under, is one type of error.

Second, if a large number of finished prints are weighed with sufficient care and accuracy, it will be found that at but very few factories the average is exactly 16 ounces, and that a manufacturer who has a 16-ounce average on one day may have a little more or a little less on the next day. *The difference of the average weight from 16 ounces* is the second type of error.

GENERAL CAUSES OF ERROR.

A more or less nonhomogeneous mass of butter, with variable physical properties, is printed on a variety of average commercially-perfect machines, which are adjusted by various arbitrary methods of checking against scales in various conditions of repair and accuracy, by men of various habits and accuracies of judgment.

Each of these manipulations contributes a component part to the total error, the causes of which are discussed in the following paragraphs.

PHYSICAL CONDITION OF THE BUTTER.

It has been shown [1] that so-called *pore space*, probably due to air spaces between the fat globules, exists in butter, and that under the best conditions the pore space is not uniformly distributed throughout the mass of butter. It has also been shown [2] that the moisture and salt content of butter made under similar conditions vary from day to day, and are not uniformly distributed throughout the mass of butter for a single day's churning. It is evident then that two

---

[1] Pickerill and Guthrie, Cornell University, Agr. Exp. Sta. Bul. 355, February, 1915.
[2] Guthrie and Ross, Cornell University, Agr. Exp. Sta. Bul. 336, October, 1913.

prints, each cut with exceptional care from different parts of a churning, may vary in weight because of the fact that one may have more pore space and less moisture or salt than the other. It has been calculated that a print so cut might differ from 16 ounces by as much as 0.21 ounce only once in about 1,500 prints. As a rule, therefore, the errors due to this cause are quite small, probably 0.08 or 0.10 ounce, but they still contribute to the total error, that is, the *difference of the single prints from 16 ounces.*

Large errors from this cause can be eliminated only by care in the manufacture of the butter, such as sprinkling the salt as uniformly as possible before working, by so working the butter as to secure as homogeneous a mixture as possible and by packing each print under the same amount of pressure.

In the course of the investigation some figures were secured on the errors due to variations in temperature at which butter is printed. At the same factory two boxes of butter, one of which had been in the ice box until well cooled, the other of which had just been churned, were cut on the same cutter. The following results were obtained:

Errors due to variations in temperature at which butter is printed.

Condition of butter.	Average difference between weight of each single print and average weight of all prints.	Difference between average weight of all prints and 16 ounces.
	Ounces.	*Ounces.*
Cool	0.43	0.35
Warm	.60	.18
Difference	.17	.17

These figures indicate what may happen when the operator of any type of machine cuts or prints cool butter on one day and warm butter on the next without readjusting the machine.

SCALES.

Any errors in the scales used to check up the printing machines are passed on to the prints when the machine is adjusted. In very few manufacturing establishments are conditions so conducive to corrosion of scales and weights as in a butter factory. Numerous cases have been found where the weights alone were in error by several hundredths of an ounce because of rust and dirt. Even with the most careful use of such a scale this error can not be eliminated except by the most careful preservation of both scales and weights.

In several instances the scales and weights were kept in a dry box when not in use. This practice is to be commended.

Many scales were found to be out of balance by 0.10 to 0.20 ounce. About 40 per cent of the scales tested were not within the U. S. Bureau of Standards tolerances, and in practically all cases the reason could be ascribed to the inattention of the butter manufacturer to the care and repair of his scale. No error greater than the U. S. Bureau of Standards tolerance of about 0.06 ounce should be introduced because of scales. Numerous makes of scales on the market at a cost not to exceed $10 easily meet this requirement.

METHOD OF ADJUSTING THE PRINTING MACHINE.

Some butter manufacturers are under the impression that a printer is set to give a pound when it is received from the factory and that no adjustment need be made from day to day. As a matter of fact, it is only an adjustable measure which must be checked quite often against an accurate standard, that is, the manufacturer's scale, to meet the moisture and pore space variations from day to day, or even at different times of the day, and to suit the peculiarities of the individual operator. About 10 per cent of the factories visited either had no scales for this purpose or, for various reasons, paid no attention to the adjustment of the printer. Twenty per cent adjusted according to the weight of 1 per cent or less of the total number of prints, while 31 per cent adjusted according to the weight of 1 to 5 per cent of the prints. The remainder, or 39 per cent, adjusted according to more than 5 per cent, 12 per cent weighing and adjusting each print. There is every reason to believe that 5 per cent is a sufficient number by which to adjust the printer if the weights are taken in the proper manner.

Many manufacturers set their scales at 1 pound and pass every print across the scales, changing the printer only when the prints run consistently off weight. Many pass only every fourth or fifth print over the scales; others weigh 5 or 10 pounds together; while still others weigh a 60-pound box for checking purposes. Although all these methods are very good, some are better than others. When single prints are weighed there is danger that the ordinary variation due to the errors of the machine may be mistaken for a change in the physical properties of the butter, while the weighing of a 60-pound box may fail to disclose some of the actual variations of the butter. Since this adjustment of the machine controls only the *difference of the average from 16 ounces*, it would seem advisable to weigh 5 or 10 pounds together and to make weighings frequently during the printing of each churning of butter.

Many men check their printers every time they make a single print; others, every fourth, fifth, or tenth print; some, two or three times during a churning; some, once a day, once a week, once a month, and some, about once a year.   As already stated, checking of single prints is not altogether desirable, but certainly checking at intervals of a week or more is not sufficient.   Since the physical properties of the butter may change from part to part of a single churning, it would seem well to check the printer four or five times at regular intervals while every separate churning is being printed.

An ideal method is: Weigh at least 5 per cent of the prints; weigh 5 or 10 together; and check four or five times at intervals during each churning, resetting if necessary until the proper weight is obtained.   It is believed that with such a procedure the *difference of the average weight from 16 ounces* need not be more than 0.05 or 0.06 ounce if no other cause of error is present.

INACCURACIES OF PRINTING MACHINES.

Without doubt the machine and its incorrect manipulation produce prints which have variable volumes and densities, thus causing comparatively large errors.   If a print is formed in a box and only one face is cut, the errors in printing can, with extreme care, be made less than those on machines which cut all six faces by means of wires, other conditions being equal.   Nevertheless, with a reasonable amount of care, the 6-face cutting machines can be operated with comparatively small variations.   More care, however, is required with them than with 1-face cutters if small variations only are to be produced.   The different types of machines, classified as to the number of faces necessarily cut by wires, a paddle, or other instrument in the process of making the print are briefly discussed in order to point out their most common peculiar causes of variation.

Type of Machine in Which One Face is Cut, the Box Being Fixed (Fig. 1).

Operation.—Butter is packed into a box with a lever tamper, or other means, and the excess struck off with a wooden paddle.

FIG. 1.

Adjustment.—The thickness of the print is controlled by screws in the bottom of the box.

Causes of Variation.—Irregularity in striking off the excess butter, which makes one edge of the print thicker than the other.   Striking with a worn or warped paddle, which makes one print larger than the next, unless care is taken to put the same side and worn place against the box each time.

Remedy.—Fill the box evenly with small portions of butter, and strike squarely across with a straight implement.   Any worn or warped striker paddles should be discarded.

Type of Machine in Which One Face is Cut, the Box Being Movable (Fig. 2).

Operation.—Similar to a biscuit cutter. Butter may or may not be spread in a layer. The excess butter is struck off by means of a paddle or other instrument.

Adjustment.—The thickness of the print is adjusted by means of screws in the bottom of the printer.

Causes of Variation.—Irregular filling of the box and irregular striking. The excess butter is sometimes struck off on the edge of a table or with a worn paddle or a crooked stick. Errors have been found, where these last two practices were in vogue, which were nearly twice the normal error for this type of machine.

Remedy.—Fill the box completely with considerable pressure, invert, and strike carefully with a straight-edged implement.

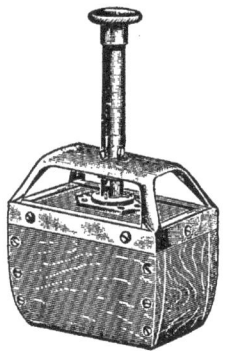

FIG. 2.

Type of Machine (Molds) in Which Two and Three Faces are Cut (Fig. 3).

Operation.—The butter is worked into a wide, shallow box which will contain 2, 3, or 4 pounds, and the excess struck off with a long paddle. The large prints

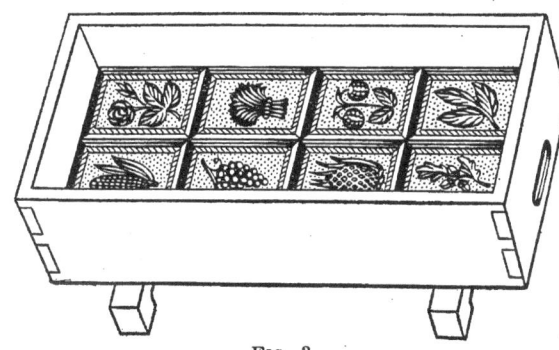

FIG. 3.

are laid side by side in a row, and 10 or 20 1-pound prints cut at one stroke with a knife.

Adjustment.—The depth of the print is regulated by screws under the bottom of the printer.

Causes of variation.—The principle of this printer is very poor. Difficulty is en-countered in securing a uniform pack in such a wide box. The surface struck is so large that if the striking paddle is raised 0.01 inch from the edge of the box an error of 0.16 ounce per pound will be made. The middles of the large prints are seldom on a line when cut.

Remedy.—Pack the butter into the box uniformly and in small portions, filling out the corners and edges, and strike with extreme care with a sharp, straight edge. Cut each large print singly on the marker line.

Type of Machine in Which Three Faces Are Cut, Commonly Called " Butter Presses " (Fig. 4).

Operation.—Tubs of butter are cut into slabs by an auxiliary machine, and placed in a beveled plunger box. The slabs are forced through an orifice, and split into bars by vertical wires. When the proper length has been forced out the plunger is stopped and several prints cut off by a wire frame.

Adjustment.—The thickness of the print is controlled by an adjustable plate in the orifice.

Causes of Variation.—Loose wires, soft butter, irregular packing in the plunger box, and insufficient attention to the distance between the cutting wires. In one case the error on one-fifth of the prints cut was practically double the error on any other print, due to a single unadjusted cutting wire. When irregular slabs are packed in the plunger box the force of the plunger is not sufficient to press out the holes in the butter, and, although the apparent holes are patched, others on the inside of the print are not seen and cause large variations.

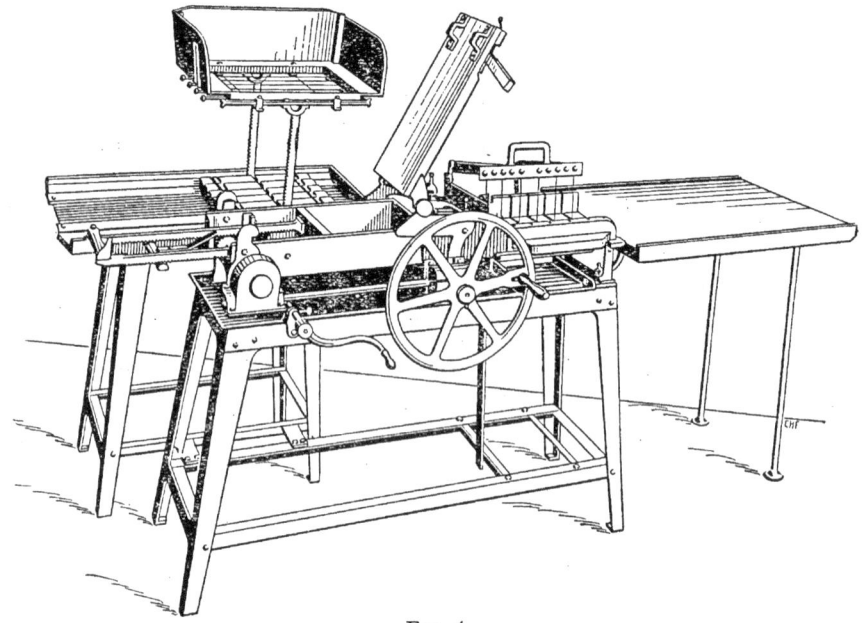

FIG. 4.

Remedy.—Carefully tighten and adjust the cutting wires at least once a week if necessary. Cut the butter as cold as consistent machine operation will allow. When cutting irregular pieces pack the plunger box solidly before closing the lid.

Type of Machine in Which Five Faces are Cut, Box Containing One Depth of Prints (Fig. 5).

Operation.—The butter is packed in a large box which will contain 25 to 30 pounds, and the excess struck off with a long rod or wire bow. The top is folded over, the box inverted, and the movable bottom pressed down, thus forcing the butter through a mesh of wire.

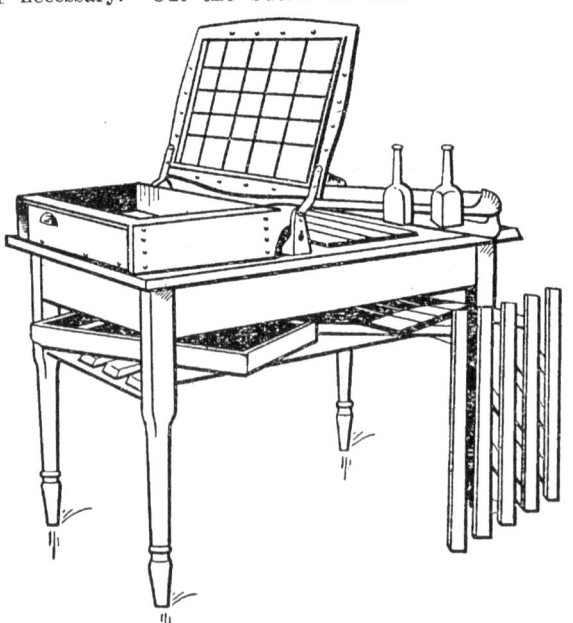

FIG. 5.

Adjustment.—The size of the print is regulated by the depth of the box.
Causes of Variation.—Irregular packing in the box and loose wires.

Remedy.—Pack uniformly, and strike with a straight instrument. Keep the wires tight and the proper distance apart.

Type of Machine in Which Six Faces are Cut, the Box Containing Several Depths of Prints (Fig. 6).

Operation.—The butter is packed in 90- to 900-pound boxes with movable bottoms, and cooled. When cold the box is placed on a piston plate, the butter forced through a wire frame, and cut off by means of a wire (cutting bow).

Adjustment.—All wires are adjustable, but variation of the width of cutting bow is the usual method.

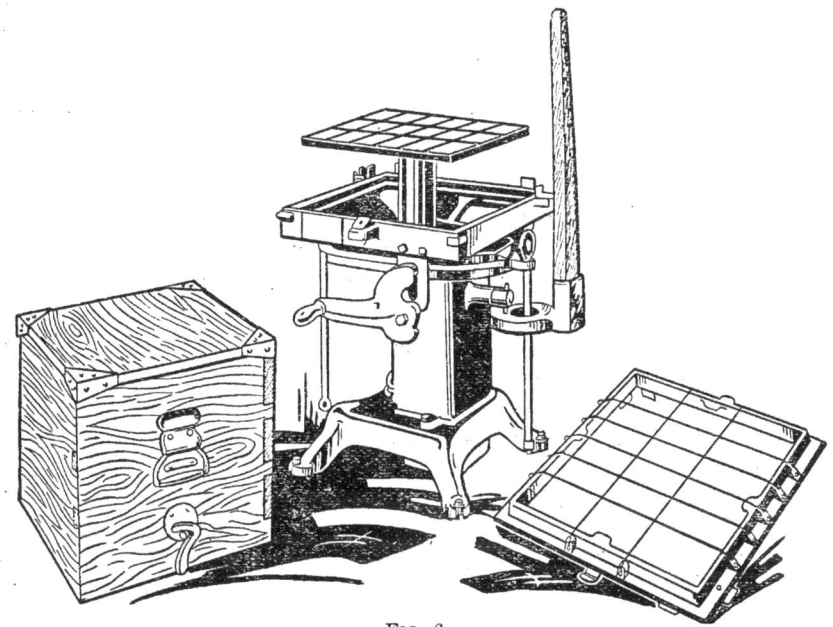

Fig. 6.

Causes of Variation.—Loose wires, irregular packing of the boxes, and careless manipulation of the bow.

Remedy.—In this, as in all other wire-cutting machines, all wires should be tight and constantly watched as to the distance apart. The butter should be so packed in the box that no air spaces are left, and every box packed to about the same solidity. The packed boxes should remain in the ice box about 24 hours before cutting to insure uniform temperature from day to day. The drawing of the bow should be done with extreme care, the measuring board kept horizontal and pressed gently against the butter.

Type of Hand-Supported Base for Adjusted Prints (Fig. 7).

Operation.—Two methods are in use: (1) The butter is first weighed out, and then

Fig. 7.

patted into a print while supported on a base similar to that shown in Figure 7; (2) it is printed on any type of machine, and then placed on the scales, and corrected or adjusted to the proper weight.

Causes of variation.—Sluggish and inaccurate scales, or careless weighing.

Remedy.—Repair or replace sluggish scales, and have all scales tested frequently by the city or State sealer. Carelessness can not be excused by the desire for speed.

Summary.

Adjustment.—All machines have adjusting devices which are very simple, so that *differences of the average from 16 ounces* can be reduced to a very small figure if sufficient care is used to check and adjust the machine.

Causes of variation.—In every case the *difference of single prints from 16 ounces* is caused by the manner of operating the machine, particularly the carelessness of the operator in manipulation, and the repair in which the machine is kept.

Remedy.—The method for the reduction of the machine errors to a minimum can be summed up in the two words, "eternal vigilance."

Investigation shows that all machines, when properly set, give about as many prints overweight as underweight, so that, in practice, *differences of the average from 16 ounces*, due to the machine alone, are not more than a very few hundredths of an ounce. Theoretically this difference should be zero. Accidents will, of course, happen occasionally, but if all suggested precautions are taken, and there is no cause for errors other than that due to the machine, it is believed that very few 16-ounce prints need be more than 0.25 ounce either above or below 16 ounces. This opinion is based on a study of the data collected.

CARELESSNESS OF THE OPERATOR.

Although the carelessness of the operator is taken up under the description of the individual printing machines (p. 8), there is another general phase which should be noted. It is a well-known fact that men differ in their accuracies of judgment. For example, in weighing on a common scale one man is inclined to give overweight, another underweight, and another to give small variations either overweight or underweight, and some to give very large variations over or under. This last type of man will cause large errors on individual prints because the "personal equation" is a very important factor in the manipulation of all printing machines. The elimination from the work or education of such men will result in much more consistent errors.

ELIMINATION OF SPECIFIC ERRORS.

DIFFERENCE OF SINGLE PRINTS FROM 16 OUNCES.

Three of the causes previously discussed contribute to this type of error, namely, (1) physical condition of the butter, (2) inaccuracies of machines, and (3) carelessness of the operator.

Butter is not a homogeneous mixture, and two prints cut from the same churning may vary in weight by an appreciable amount (probably 0.10 ounce from the intended weight), even if cut with the greatest care and exactly the same size. The machines are not perfect, and, even with perfect repairs and extremely careful operation on an entirely homogeneous product, might make an error of about 0.25 ounce. Field experience teaches that machines which cut a large number of faces of the print are more often out of repair than those cutting fewer faces, and many large errors can be attributed to this fact. An inefficient operator introduces many causes of errors, such as striking on the edge of a table, or, with a canted paddle, failure to pack the butter solidly before cutting, or by his natural inability to repeat identical operations. No estimate can be made as to the amount of error such an operator might introduce.

This type of error may be reduced to a minimum by careful manufacture of the butter to secure a homogeneous product, careful manipulation, and daily attention to the repair of the printing or cutting machine, and the selection or education of an operator.

DIFFERENCE OF THE AVERAGE WEIGHT FROM 16 OUNCES.

This type of error is contributed to by the remaining two causes previously discussed, namely, (1) inaccurate scales, and (2) incorrect methods of adjusting the printing machine. No machine can be more perfect than the scale against which it is checked. It is therefore imperative that the butter manufacturer have a scale of commercially practicable accuracy. Such a scale will sometimes have an inaccuracy in itself of about 0.06 ounce per pound according to the U. S. Bureau of Standards system of tolerances. There is, however, no reason why prints should be off more than 0.06 ounce from this cause. The variety of inefficient methods in vogue for checking the machine suggests that there is vast room for improvement in this most important and most neglected step of the printing process. Enough prints must be weighed every day to make sure that the average has been found before the printer is reset. Under the most efficient methods it is estimated that an error of 0.05 or 0.06 ounce is probable.

This type of error may be reduced to a minimum by first securing a good commercial scale, and then preserving it to insure its accuracy in the future, and weighing at least 5 per cent of all prints, either singly or 5 or 10 together, checking four or five times at intervals during each churning and resetting if necessary until the proper weight is obtained.

SUMMARY.

Investigations show that erroneous opinions exist as to the magnitude of the errors produced by present commercial manipulations of butter printing and cutting devices. Attention to many details is necessary in order to produce only errors as small as those required by the simple provisions of the Federal net weight amendment and the regulations thereunder.

Errors on single prints, when the average is correct, are due principally to the physical condition of the butter, the inaccuracy of the printing machine, and the carelessness of the operator. These errors may be largely eliminated by attention to such details as uniformity of mixing, control of the temperature at which printed, securing a uniform solidity of the print, cutting all prints squarely, filling out the corners, preventing air holes in the middle of the print, elimination of worn utensils, and keeping cutting wires tight and the proper distance apart.

Errors on the average weight are due largely to inaccurate scales and incorrect methods of adjusting the machine. They may be largely eliminated by first securing an accurate scale and then looking carefully to its preservation; also by weighing at least 5 per cent of all prints in groups of five or ten at frequent intervals during each churning, in order to check up the printing machine.

Actual conditions as they were found in the field are discussed in this circular, and the relation of the different sources of error to one another is pointed out.

PUBLICATIONS OF THE U. S. DEPARTMENT OF AGRICULTURE RELATING TO BUTTER.

AVAILABLE FOR FREE DISTRIBUTION BY THE DEPARTMENT.

Making Butter on the Farm. (Farmers' Bulletin 876.)

Marketing Creamery Butter. (Department Bulletin 456.)

Suggestions for Manufacture and Marketing of Creamery Butter in the South. (Secretary's Circular 66.)

Butter Making in the South. (Secretary's Special.)

FOR SALE BY THE SUPERINTENDENT OF DOCUMENTS, GOVERNMENT PRINTING OFFICE, WASHINGTON, D. C.

Household Tests for Detection of Oleomargarine and Renovated Butter. (Farmers' Bulletin 131.) Price, 5 cents.

Farm Butter Making. (Farmers' Bulletin 541.) Price, 5 cents.

Production and Consumption of Dairy Products. (Department Bulletin 177.) Price, 5 cents.

Studies Upon Keeping Quality of Butter: 1 Canned Butter. (Bureau of Animal Industry Bulletin 57.) Price, 5 cents.

Investigations in Manufacture and Storage of Butter: 1 Keeping Qualities of Butter Made Under Different Conditions and Stored at Different Temperatures with Remarks on Scoring of Butter. (Bureau of Animal Industry Bulletin 84.) Price, 10 cents.

Influence of Acidity of Cream on Flavor of Butter. (Bureau of Animal Industry Bulletin 114.) Price, 10 cents.

Normal Composition of American Creamery Butter. (Bureau of Animal Industry Bulletin 149.) Price, 5 cents.

Factors Influencing Change in Flavor in Storage Butter. (Bureau of Animal Industry Bulletin 162.) Price, 10 cents.

Paraffining Butter Tubs. (Bureau of Animal Industry Circular 130.) Price, 5 cents.

Fishy Flavor in Butter. (Bureau of Animal Industry Circular 146.) Price, 5 cents.

Whey Butter. (Bureau of Animal Industry Circular 161.) Price, 5 cents.

Simple Butter Color Standard. (Bureau of Animal Industry Circular 200.) Price, 5 cents.

CPSIA information can be obtained
at www.ICGtesting.com
Printed in the USA
BVHW090434201118
533516BV00014B/964/P

9 781528 024440